ENERGY SECTOR STANDARD OF THE PEOPLE'S REPUBLIC OF CHINA

中华人民共和国能源行业标准

Specification for Preparation of Feasibility Study Report on Densified Biofuel Heating Projects

生物质成型燃料供热工程
可行性研究报告编制规程

NB/T 34039-2017

Chief Development Department: China Renewable Energy Engineering Institute
Approval Department: National Energy Administration of the People's Republic of China
Implementation Date: August 1, 2017

China Water & Power Press
中国水利水电出版社
Beijing 2024

All rights reserved. No part of this publication may be reproduced, stored in a retrieval system, or transmitted in any form or by any means—electronic, mechanical, photocopying, recording or otherwise, without prior written permission of the publisher.

图书在版编目（CIP）数据

生物质成型燃料供热工程可行性研究报告编制规程：NB/T 34039-2017 = Specification for Preparation of Feasibility Study Report on Densified Biofuel Heating Projects (NB/T 34039-2017)：英文 / 国家能源局发布. -- 北京：中国水利水电出版社, 2024. 7. ISBN 978-7-5226-2647-5

Ⅰ. TK63-65

中国国家版本馆CIP数据核字第2024Y08U36号

ENERGY SECTOR STANDARD
OF THE PEOPLE'S REPUBLIC OF CHINA
中华人民共和国能源行业标准

Specification for Preparation of Feasibility Study Report
on Densified Biofuel Heating Projects
生物质成型燃料供热工程可行性研究报告编制规程
NB/T 34039-2017
（英文版）

Issued by National Energy Administration of the People's Republic of China
国家能源局　发布
Translation organized by China Renewable Energy Engineering Institute
水电水利规划设计总院　组织翻译
Published by China Water & Power Press
中国水利水电出版社　出版发行
　　Tel: (+ 86 10) 68545888　68545874
　　sales@mwr.gov.cn
　　Account name: China Water & Power Press
　　Address: No.1, Yuyuantan Nanlu, Haidian District, Beijing 100038, China
　　http://www.waterpub.com.cn
中国水利水电出版社微机排版中心　排版
北京中献拓方科技发展有限公司　印刷
184mm×260mm　16开本　4.5印张　142千字
2024年7月第1版　2024年7月第1次印刷
Price（定价）：￥740.00

Introduction

This English version is one of China's energy sector standard series in English. Its translation was organized by China Renewable Energy Engineering Institute authorized by National Energy Administration of the People's Republic of China in compliance with relevant procedures and stipulations. This English version was issued by National Energy Administration of the People's Republic of China in Announcement [2021] No. 3 dated April 26, 2021.

This version was translated from the Chinese Standard NB/T 34039-2017, *Specification for Preparation of Feasibility Study Report on Densified Biofuel Heating Projects*, published by China Electric Power Press. The copyright is reserved by National Energy Administration of the People's Republic of China. In the event of any discrepancy in the implementation, the Chinese version shall prevail.

Many thanks go to the staff from the relevant standard development organizations and those who have provided generous assistance in the translation and review process.

For further improvement of the English version, any comments and suggestions are welcome and should be addressed to:

China Renewable Energy Engineering Institute
No. 2 Beixiaojie, Liupukang, Xicheng District, Beijing 100120, China
Website: www.creei.cn

Translating organizations:

POWERCHINA Northwest Engineering Corporation Limited

China Renewable Energy Engineering Institute

Translating staff:

LI Zhongjie	QIAO Peng	LI Kejia	ZHANG Peng
MA Gaoxiang	XU Chenchen		

Review panel members:

QIE Chunsheng	Senior English Translator
YAN Wenjun	Army Academy of Armored Forces, PLA
JIN Feng	Tsinghua University
LI Angui	Xi'an University of Architecture and Technology

LIU Xiaofen	POWERCHINA Zhongnan Engineering Corporation Limited
CHEN Lei	POWERCHINA Zhongnan Engineering Corporation Limited
JIA Haibo	POWERCHINA Kunming Engineering Corporation Limited
XIE Hongwen	China Renewable Energy Engineering Institute
LI Shisheng	China Renewable Energy Engineering Institute

National Energy Administration of the People's Republic of China

翻译出版说明

本译本为国家能源局委托水电水利规划设计总院按照有关程序和规定，统一组织翻译的能源行业标准英文版系列译本之一。2021年4月26日，国家能源局以2021年第3号公告予以公布。

本译本是根据中国电力出版社出版的《生物质成型燃料供热工程可行性研究报告编制规程》NB/T 34039—2017翻译的，著作权归国家能源局所有。在使用过程中，如出现异议，以中文版为准。

本译本在翻译和审核过程中，本标准编制单位及编制组有关成员给予了积极协助。

为不断提高本译本的质量，欢迎使用者提出意见和建议，并反馈给水电水利规划设计总院。

    地址：北京市西城区六铺炕北小街2号
    邮编：100120
    网址：www.creei.cn

本译本翻译单位：中国电建集团西北勘测设计研究院有限公司
       水电水利规划设计总院

本译本翻译人员：李仲杰 乔 鹏 李可佳 张 鹏
       马高祥 许晨琛

本译本审核人员：

    郄春生 英语高级翻译

    闫文军 中国人民解放军陆军装甲兵学院

    金 峰 清华大学

    李安桂 西安建筑科技大学

    刘小芬 中国电建集团中南勘测设计研究院有限公司

    陈 蕾 中国电建集团中南勘测设计研究院有限公司

    贾海波 中国电建集团昆明勘测设计研究院有限公司

    谢宏文 水电水利规划设计总院

    李仕胜 水电水利规划设计总院

国家能源局

Announcement of National Energy Administration of the People's Republic of China
[2017] No. 6

According to the requirements of Document GNJKJ [2009] No. 52, "Notice on Releasing the Energy Sector Standardization Administration Regulations (*tentative*) and detailed implementation rules issued by National Energy Administration of the People's Republic of China", 159 sector standards such as *Shale Gas Reservoir Stimulation—Part 2: Technical Specification for Factory Fracturing Operation*, including 34 energy standards (NB), 39 electric power standards (DL), and 86 petroleum standards (SY), are issued by National Energy Administration of the People's Republic of China after due review and approval.

The electric power standards are published by China Electric Power Press or China Planning Press; coal standards by China Coal Industry Publishing House; petroleum, natural gas and shale gas standards by Petroleum Industry Press; and boiler and pressure vessel standards by Xinhua Publishing House.

Attachment: Directory of Sector Standards

National Energy Administration of the People's Republic of China

March 28, 2017

Attachment:

Directory of Sector Standards

Serial number	Standard No.	Title	Replaced standard No.	Adopted international standard No.	Approval date	Implementation date
...						
12	NB/T 34039-2017	Specification for Preparation of Feasibility Study Report on Densified Biofuel Heating Projects			2017-03-28	2017-08-01
...						

Foreword

According to the requirements of Document GNKJ [2015] No. 12 issued by National Energy Administration of the People's Republic of China, "Notice on Releasing the Development and Revision Plan of the Second Batch of Energy Sector Standards in 2014", and after extensive investigation and summarization of research and practical experience in design, manufacturing, construction and operation management in biomass energy sector in China, consultation of relevant standards at home and abroad, and wide solicitation of opinions, the drafting group has prepared this specification.

The main technical contents of this specification include: preparation basis and tasks, executive summary, necessity and feasibility of construction, heating load, densified biofuel supplies, project scale, properties of densified biofuel and boiler selection, project site selection, general scheme, civil works, electrical, monitoring, control, and utility system, fire protection, organization and staffing, project scheduling, environmental protection, energy conservation and emission reduction, occupational health and safety, cost estimate, financial evaluation and social benefit analysis, project bidding, conclusions and suggestions, attachments, and drawings.

National Energy Administration of the People's Republic of China is in charge of the administration of this specification. China Renewable Energy Engineering Institute has proposed this specification, is responsible for its routine management, and is responsible for the explanation of specific technical contents. Comments and suggestions in the implementation of this specification should be addressed to:

China Renewable Energy Engineering Institute
No. 2 Beixiaojie, Liupukang, Xicheng District, Beijing 100120, China

Chief development organizations:

POWERCHINA Northwest Engineering Corporation Limited

China Renewable Energy Engineering Institute

Participating development organizations:

China Architecture Design & Research Group

Academy of Agricultural Planning and Engineering

Xi'an University of Architecture and Technology

Chief drafting staff:

ZHANG Peng	YANG Zhigang	FENG Tao	ZHENG Kun
NIU Wenbin	HU Xiaofeng	CHEN Wei	XU Chenchen
ZHUANG Kun	MA Bo	LIU Xuan	ZHAO Lixin
MENG Haibo	YAO Zonglu	WANG Jilin	XU Junjie
LI Angui			

Review panel members:

XIE Hongwen	JIA Zhenhang	DOU Kejun	YU Guosheng
FENG Yongzhong	BIAN Zhimin	MENG Wei	YU Jiaqing
TIAN Xiaoxia	WANG Biao	LI Shujun	LIU Qijun
DONG Zhou	LIU Yuzhuo	YE Yang	LI Shisheng

Contents

1	**General Provisions** ··	1
2	**Terms** ···	2
3	**Preparation Basis and Tasks** ··	3
3.1	Preparation Basis ··	3
3.2	Basic Data ··	3
3.3	Preparation Tasks ···	3
4	**Executive Summary** ··	6
5	**Necessity and Feasibility of Construction** ···························	8
5.1	Project Background ··	8
5.2	Necessity of Construction ···	8
5.3	Feasibility of Construction ··	8
6	**Heating Load** ···	10
6.1	Present Heating Status ···	10
6.2	Heating Load ···	10
6.3	Design Heating Load ··	13
6.4	Heating Load Diagram ···	14
7	**Densified Biofuel Supplies** ···	15
7.1	Market Demand Analysis ···	15
7.2	Supply and Security ···	15
8	**Project Scale** ···	16
9	**Properties of Densified Biofuel and Boiler Selection** ········	17
9.1	Properties of Densified Biofuel ··	17
9.2	Boiler Selection ···	17
10	**Project Site Selection** ···	18
10.1	Site Selection ···	18
10.2	Site Overview ··	18
10.3	Engineering Geology ··	18
10.4	Traffic Conditions ···	18
10.5	Water and Power Supply ··	19
11	**General Scheme** ··	20
11.1	Project Planning and General Layout of Heating Plant ······	20
11.2	Overall Design of Heating Plant ··	21
11.3	Overall Design of Heating Pipe Network ····························	22
12	**Civil Works, Electrical, Monitoring, Control and Utility System** ···	25
12.1	Civil Works ··	25
12.2	Electrical ··	25

12.3	Monitoring and Control	26
12.4	Heating, Ventilation, and Air Conditioning	26
12.5	Water Supply and Drainage	26
13	**Fire Protection**	**28**
13.1	Overall Design of Fire Protection	28
13.2	Fire Protection Engineering Design	28
14	**Organization and Staffing**	**30**
14.1	Project Management Organization	30
14.2	Main Management Facilities	30
15	**Project Scheduling**	**31**
15.1	Project Implementation Conditions	31
15.2	Construction Schedule	31
16	**Environmental Protection, Energy Conservation and Emission Reduction**	**32**
16.1	Environmental Protection	32
16.2	Energy Conservation and Emission Reduction	32
17	**Occupational Health and Safety**	**34**
17.1	Identification and Analysis of Main Hazard Factors	34
17.2	Engineering Design of Occupational Health and Safety	34
17.3	Safety Management During Operation	35
17.4	Special Investment for Occupational Health and Safety	35
18	**Cost Estimate**	**36**
18.1	Preparation Instructions	36
18.2	Cost Estimate Table	36
19	**Financial Evaluation and Social Benefit Analysis**	**37**
19.1	Overview	37
19.2	Financial Evaluation	37
19.3	Social Benefit Evaluation	38
19.4	Financial Evaluation Table	38
20	**Project Bidding**	**39**
20.1	Bidding Scope	39
20.2	Bidding Method and Organization	39
21	**Conclusions and Suggestions**	**40**
21.1	Conclusions	40
21.2	Suggestions	40
22	**Attachments and Drawings**	**41**
22.1	Attachments	41
22.2	Drawings	41
Appendix A	**Basic Data for Preparation of Feasibility Study Report on Densified Biofuel Heating Projects**	**43**

Appendix B　Contents of Preparation of Feasibility Study
　　　　　　Report on Densified Biofuel Heating Projects ··· 47
Appendix C　Main Techno-Economic Index of Densified Biofuel
　　　　　　Heating Projects ··· 50
Explanation of Wording in This Specification ···················· 54
List of Quoted Standards ··· 55

1 General Provisions

1.0.1 This specification is formulated with a view to standardizing the preparation of feasibility study report on densified biofuel heating projects.

1.0.2 This specification is applicable to the construction, renovation and extension of densified biofuel heating projects with a rated evaporation of 10 t/h or a rated thermal power of 7 MW and above for single boiler.

1.0.3 Prior to the feasibility study on densified biofuel heating projects, the agriculture and forestry biomass resources shall be investigated and evaluated, and a report shall be prepared.

1.0.4 In addition to this specification, preparation of the feasibility study report on densified biofuel heating projects shall comply with other current relevant standards of China.

2 Terms

2.0.1 densified biofuel

solid biofuel made by mechanically compressing biomass into a specific size and shape

2.0.2 economic supply radius

mean distance from the heating plant to the farthest densified biofuel suppliers

2.0.3 heating plant

heating complex using boiler as heat source

2.0.4 heating system with multi-heat sources

heating system which has multiple heat sources, and three operating modes, namely, the independent operation, separate operation, and pooled operation of multi-heat sources

3 Preparation Basis and Tasks

3.1 Preparation Basis

The feasibility study report on a densified biofuel heating project shall be prepared in accordance with the national biomass energy policies, laws and regulations, and technical standards.

3.2 Basic Data

3.2.1 Basic data shall include densified biofuel heating project planning and preparatory work data.

3.2.2 The feasibility study shall thoroughly investigate the construction conditions of densified biofuel heating project and obtain reliable basic data, which shall meet the requirements of Appendix A of this specification.

3.2.3 Socioeconomic status and development plan, land use plan and environmental constraints of the region where the project is located shall be collected.

3.2.4 The data on local agriculture and forestry resources and their utilization shall be collected.

3.2.5 The data on physical geography, accessibility and firefighting facilities in the region where the project is located shall be collected.

3.2.6 The prices of main construction materials and the policies and regulations concerning construction cost for the region where the project is located shall be collected.

3.2.7 The state and local financial and tax incentives that the project can enjoy shall be collected.

3.3 Preparation Tasks

3.3.1 The feasibility study report on the densified biofuel heating project shall:

1. Describe project background; demonstrate the project necessity and feasibility.

2. Present local heating supply status; collect and check the local heating load data; determine design heating load and heating load diagram; forecast local medium- and long-term heating load growth.

3. Describe the local agriculture and forestry biomass resources and utilization status based on investigation and evaluation; analyze the densified biofuel market supply condition, and present evaluation

conclusion.

4 Determine project tasks and scale.

5 Determine boiler type and model through techno-economic comparison according to the type, nature and performance index of densified biofuel locally available.

6 Describe the conditions for project site selection and put forward the proposed site with due consideration of engineering geology, fuel supply, transport, water and power supply, surroundings, etc.

7 Determine the overall design scheme of heating project, including handling and storage of densified biofuel in heating plant, biofuel conveying and feeding system, boiler thermal system, boiler flue gas and air system, slag removing and ash storage devices, boiler water treatment and water supply systems, thermotechnical monitoring and control systems; heating medium and parameters of heating network, type and laying of heating pipe network; connection between heat supply station, heating pipe network and users; heating network regulation, dispatching and monitoring and control; heat metering; hydraulic calculation for pipeline; condensate recovery in pipe network, etc.

8 Describe the design schemes for civil works, electrical, lighting, earthing and lightning protection, monitoring and control, heating and ventilation and air conditioning (HVAC), and water supply and drainage.

9 Propose the fire protection design scheme.

10 Propose the project organizational structure and staffing and put forward the main management facilities.

11 Put forward the critical path for the construction schedule and the key factors that affect the progress of the main works according to the implementation condition of the project.

12 Propose the environmental protection, energy conservation and emission reduction design.

13 Propose the occupational health and safety scheme.

14 Propose the cost estimate.

15 Make financial evaluation and social benefit analysis.

16 Propose the project bidding scheme.

17　Present conclusions and suggestions for project construction.

3.3.2　The contents of the feasibility study report on the densified biofuel heating project should meet the relevant requirements in Appendix B of this specification.

4 Executive Summary

4.0.1 Executive summary should include: project necessity and feasibility; heating load; densified biofuel supplies; project scale; boiler type selection; project site selection; overall design scheme; civil works, electrical, monitoring and control, and utility system; fire protection; staffing; project schedule; environmental protection, energy conservation and emission reduction; occupational health and safety; cost estimate; financial evaluation and social benefit analysis; project bidding; conclusions and suggestions; and a brief description of the project investor(s) and the project owner.

4.0.2 The project necessity and feasibility shall be briefly stated in terms of project background, construction conditions, environment, socioeconomic benefits, etc.

4.0.3 The local heating source distribution, heating mode, and heating network situation shall be briefly described. The current heating load, planned heating load and design heating load shall be proposed based on heating load investigation and verification.

4.0.4 The main conclusions of the investigation and evaluation on agriculture and forestry biomass resources in and around the project area shall be presented; and the production, collection, transportation, storage and supply of densified biofuel in the project area should be briefly described.

4.0.5 Project tasks and scale shall be proposed according to regional economic development status and short- and long-term development plans, the status quo and development plan of district heating and heating source system, etc. The main factors affecting the project scale and the demonstration process of the final scale should be briefly described.

4.0.6 Boiler type and model shall be proposed through techno-economic comparison of alternatives according to densified biofuel nature, heating load characteristics, biofuel combustion, construction and operation costs, etc.

4.0.7 Project site and land use scheme shall be proposed. The geographical location, natural condition, surroundings, engineering geology, accessibility, and water and power supply for the proposed site shall be briefly described.

4.0.8 The overall design for heating plant and heating pipe network shall be briefly described.

4.0.9 The general layout, plan, architectural design, structure type, and foundation treatment scheme for main buildings (structures), and the flood control, waterlogging protection, earthquake-proof and explosion-proof facilities shall be briefly described. Special measures to allow the pipe network passing through or spanning roads, rivers and other obstacles should be briefly described. The power load and supply mode shall be briefly described. A brief description of power supply and distribution, artificial lighting, lightning protection and earthing, monitoring and control and communication system design shall be provided. HVAC and water supply and drainage system design should be briefly described.

4.0.10 The overall design of fire protection and main fire-fighting facilities shall be briefly described.

4.0.11 Project management organization and main management facilities should be briefly described.

4.0.12 Project implementation conditions and construction schedule should be briefly described.

4.0.13 Environmental protection design shall be briefly described, the main energy consumption types, quantity, and energy conservation during operation period should be put forward; main energy conservation measures and energy conservation and emission reduction benefits should be outlined.

4.0.14 The main conclusions of the safety pre-assessment report should be provided, and the main hazards shall be briefly described. The main design content and investment for occupational health and safety shall be briefly described.

4.0.15 Total investment, static investment, investment per kW heat output, investment components and fundraising plan shall be briefly described.

4.0.16 The main results and conclusions of financial evaluation and social benefit analysis shall be outlined.

4.0.17 The scope, method and organization of bidding should be briefly described.

4.0.18 The main conclusions on project construction shall be provided, and recommendations for further work should be put forward.

4.0.19 Table of main techno-economic indicators and summary table of cost estimate for the densified biofuel heating project shall be prepared. The main techno-economic indicators of the densified biofuel heating project shall be in accordance with Appendix C of this specification.

5 Necessity and Feasibility of Construction

5.1 Project Background

5.1.1 Project background should state the project purposes as follows:

1. Improve the ecological environment, and develop low-carbon recycling economy.

2. Improve the utilization rate of agriculture and forestry biomass resources, optimize energy consumption structure and maintain energy security.

3. Achieve energy conservation and emission reduction, and tackle climate change.

4. Strengthen the prevention and control of atmospheric pollution and haze governance, and guide the comprehensive utilization of agriculture and forestry biomass resources.

5. Promote the improvement of environmental hygiene in villages and towns, and improve the quality of life and health of local residents.

5.2 Necessity of Construction

5.2.1 The state industrial policies regarding coping with climate change, promoting the prevention and control of air pollution, promoting energy structure transformation and strengthening environmental protection, as well as the state renewable energy and biomass energy development status and planning shall be outlined, and the necessity of heating utilization and development of biomass resources shall be analyzed and demonstrated in terms of the compliance with the state industrial policies and planning, building a heating system with renewable energy and promoting new urbanization.

5.2.2 The status of local energy consumption shall be briefly described, and the necessity of project construction shall be demonstrated from the perspective of rational utilization of biomass resources.

5.2.3 The effect of construction of the densified biofuel heating project on promoting local socio-economic development shall be analyzed.

5.2.4 The construction conditions, and environmental and economic benefits of the densified biofuel heating project shall be outlined, to demonstrate the necessity of project construction.

5.3 Feasibility of Construction

5.3.1 The technological feasibility of the project shall be initially analyzed with due consideration of factors such as energy conversion efficiency, the guarantee degree of agriculture and forestry biomass resources and densified biofuel, technology maturity, process reliability, system energy and water consumption, market share, and development potential.

5.3.2 The economic rationality of the project shall be initially analyzed in terms of project investment, comprehensive cost, operation and management expenses, rate of return, and payback period of investment.

5.3.3 The uncertainties and risks of the densified biofuel heating project shall be initially analyzed.

5.3.4 The feasibility of project construction shall be demonstrated through comprehensive comparison of densified biofuel with coal, oil, gas or other different fuels in terms of environmental protection and cost effectiveness, taking into account the results in Articles 5.3.1 to 5.3.3 of this specification, and following the principles that the project meets the user demand, has sufficient raw materials supply and is technically feasible.

6 Heating Load

6.1 Present Heating Status

6.1.1 The local heating source distribution and heating present status, heating mode and heating network shall be described; the number of existing heating boilers, annual coal (oil, gas) consumption, and the environmental pollution status within the heating scope shall be elaborated.

6.2 Heating Load

6.2.1 The investigation and verification results of existing heat consumption enterprises and existing heating building area shall be presented in accordance with the following requirements:

1. The heating load for existing heat consumption enterprises within heating scope obtained from the investigation shall be listed. The summary heating load for existing heat consumption enterprises should be in accordance with Table 6.2.1.

2. The investigation results of domestic heating load shall be provided, the existing building area within the heating scope of heating plant shall be described according to building type, and the number and capacity of heating boilers shall be listed; the existing maximum, average and minimum space-heating load are calculated according to outdoor design temperature for local heating and mean temperature in heating period, duration of heating period, heat index for heat supply, and existing heating area; the maximum, average and minimum heating load for cooling are calculated according to existing air conditioners' cooling area, duration of the cooling period, and heat index for cooling; the heating load for domestic hot water shall be calculated according to the existing building area with access to domestic hot water and heat index.

6.2.2 The investigation and verification of the near-term heating load shall meet the following requirements:

1. The near-term industrial heating load shall include the existing industrial heating load and the industrial heating load increase in the near term.

2. The near-term domestic heating load shall include the existing domestic heating load and the domestic heating load increase in the near term.

6.2.3 In heating load calculation, the planned projects and urban development plan under the heating coverage of the project shall also be considered in terms of industrial and civil purposes.

Table 6.2.1　Summary heating load for existing heat consumption enterprises

S/N	Enterprise	Product name and output	Existing boiler		Heating medium temperature (°C)	Heating medium pressure (MPa)	Production shift (shift/d)	Number of annual production days (d)	Load in heating period (t/h)			Load in cooling period (t/h)			Load in non-heating, non-cooling period (t/h)		
			Capacity (t/h or MW)	Qty.					Max.	Mean	Min.	Max.	Mean	Min.	Max.	Mean	Min.

Table 6.3.1　Industrial heating load summary

S/N	Enterprise	Heating medium pressure (MPa)	Heating medium temperature (°C)	Production shift (shift/d)	Existing heating load (t/h)											Near-term heating load (t/h)											Production days (d)
					Heating period			Non-heating period			Non-heating, non-cooling period			Heating period			Non-heating period			Non-heating, non-cooling period			
					Max.	Mean	Min.	Max.	Mean	Min.	Max.	Mean	Min.	Max.	Mean	Min.	Max.	Mean	Min.	Max.	Mean	Min.	

6.3 Design Heating Load

6.3.1 After enthalpy conversion of the verified industrial heating load, an industrial heating load summary shall be provided, which should be in accordance with Table 6.3.1.

6.3.2 The space-heating load summary should be in accordance with Table 6.3.2.

Table 6.3.2　Space-heating load summary

District	Area (m^2)	Heat index (W/m^2)	Heating load (GJ/h)
…	…	…	…
Total			

6.3.3 The heating load summary for cooling by air conditioning should be in accordance with Table 6.3.3.

Table 6.3.3　Heating load summary for cooling by air conditioning

District	Area (m^2)	Heat index (W/m^2)	Heating load (GJ/h)
…	…	…	…
Total			

6.3.4 The heating load summary for domestic hot water should be in accordance with Table 6.3.4.

Table 6.3.4　Heating load summary for domestic hot water

District	Area (m^2)	Heat index (W/m^2)	Heating load (GJ/h)
…	…	…	…
Total			

6.3.5 The return condition of condensate water of heating shall be described, including water quality, return rate, recovery methods.

6.3.6 Considering heating network loss and the simultaneity coefficient for maximum steam consumption in industrial enterprises, the design heating load table shall be worked out. The design heating load should be in accordance with Table 6.3.6.

Table 6.3.6 Design heating load (t/h or MW)

Category	Heating period			Cooling period			Non-heating, non-cooling period		
	Max.	Mean	Min.	Max.	Mean	Min.	Max.	Mean	Min.
Industry									
Space heating									
Cooling by air conditioning									
Domestic hot water									
Total									

6.4 Heating Load Diagram

6.4.1 According to the heat consumption of existing enterprises in Table 6.3.1 of this specification, annual production heating load curves shall be plotted by production shift system and the number of annual production days.

6.4.2 The space-heating (air conditioning) load curves shall be plotted, including hourly space-heating (air conditioning) load curve, annual space-heating (air conditioning) load continuation curve, etc.

6.4.3 Heating load curve for domestic hot water supply should be plotted.

7 Densified Biofuel Supplies

7.1 Market Demand Analysis

7.1.1 Densified biofuel market analysis, including market capacity prediction and price prediction, shall be conducted based on the resources survey and evaluation for agriculture and forest biomass heating projects.

7.1.2 Market capacity shall be predicted according to proposed alternative fuels, densified biofuel users, and densified biofuel demand.

7.1.3 Predictive analysis shall be conducted on densified biofuel product prices according to raw material acquisition prices, fuel processing costs, transport costs, the price level of similar products, local acceptance level, etc.

7.2 Supply and Security

7.2.1 The production process, scale, average annual production capacity and geographical location of densified biofuel enterprises (suppliers) shall be described.

7.2.2 The acquisition, transportation, and storage conditions of densified biofuel shall be briefly described, market supply analysis shall be conducted on densified biofuel, and market supportability of densified biofuel shall be explained.

7.2.3 The economic supply radius of densified biofuel shall be reasonably determined.

8 Project Scale

8.0.1 Principal factors affecting the scale of the densified biofuel heating project shall be analyzed, such as fuel costs, fixed asset investment, heating load, and heat service coverage.

8.0.2 The project scale shall be demonstrated and determined according to the densified biofuel resources in the region where the project is located, district heating status and development plan, project construction conditions, etc., taking into account the status of biomass energy utilization technology and the equipment research, development and manufacturing level.

8.0.3 For a project developed in phases, a brief description may be given according to the scale and scope of the project in other phases.

9 Properties of Densified Biofuel and Boiler Selection

9.1 Properties of Densified Biofuel

9.1.1 A testing report of densified biofuel issued by a qualified testing organization shall be provided.

9.1.2 The type, physical properties, ultimate analysis, proximate analysis, calorific value, combustion characteristics, clinkering property and pollutant property of the densified biofuel used by heating plant shall be briefly described.

9.2 Boiler Selection

9.2.1 The various technical indicators of selected boiler shall meet the requirements of the current national standards GB 24500, *The Minimum Allowable Values of Energy Efficiency and Energy Efficiency Grades of Industrial Boilers,* and GB/T 15317, *Monitoring and Testing for Energy Saving of Coal Fired Industrial Boilers*, and the following factors shall be taken into account:

1 Nature and different combustion technologies of densified biofuel.

2 Heating load characteristics.

3 The impact of the ash deposit, clinkering and corrosion of the minerals in densified biofuel on heating efficiency.

4 The mechanization and automation of the boiler.

5 Environmental protection requirements.

6 Construction cost and operation cost.

9.2.2 Boiler models and specifications shall be recommended through techno-economic comparison.

10 Project Site Selection

10.1 Site Selection

10.1.1 A techno-economic comparison shall be conducted on two or more site alternatives according to the economic supply radius of local densified biofuel, heating load concentration ratio and the general conditions of heating plant construction, and the recommended plant site shall be put forward.

10.2 Site Overview

10.2.1 A brief introduction shall be given to the administrative district and geographical location of the heating plant.

10.2.2 The topography, geomorphology, engineering geology, mineral resources, hydrology, meteorology, etc. in the region where the heating plant site is located shall be outlined.

10.2.3 The flood control standard for the plant site area shall be given. When the plant site is likely to be affected by waterlogging, the waterlogging level shall be analyzed.

10.2.4 The relationship between heating plant site and urban and rural planning, and surrounding major rivers, roads, railways, waterways, and nature reserves, cultural heritage reserves, places of historic interest and scenic beauty, airports, military installations, communication facilities, etc. shall be described.

10.3 Engineering Geology

10.3.1 An engineering geological investigation or relevant geological data collection of selected plant site shall be carried out. A preliminary scheme for foundation selection and ground treatment of the major buildings (structures) of the heating plant shall be proposed according to geological conditions.

10.3.2 The hazard rating, distribution range, and development trend of the unfavorable geological phenomenon which endangers plant site shall be identified, and a treatment scheme shall be put forward.

10.3.3 The basic seismic intensity and peak ground acceleration at the project site shall be determined in accordance with the current national standard GB 18306, *Seismic Ground Motion Parameters Zonation Map of China*.

10.3.4 The plant site and groundwater source shall not be located in a mine with exploitation value.

10.4 Traffic Conditions

10.4.1 The status quo and planning of the roads near the plant site shall be briefly described.

10.4.2 The transportation modes and conditions such as roads, railways and waterways shall be briefly described. The possible transportation modes shall be discussed and the recommended scheme shall be put forward according to the situations of fuel supply center, temporary storage station, transfer station, transport volume, and traffic conditions.

10.5 Water and Power Supply

10.5.1 The kinds of water sources shall be determined; water withdrawal mode and water supply system scale shall be selected.

10.5.2 Electric power sources, voltage level, and power transmission and transformation schemes shall be proposed.

11 General Scheme

11.1 Project Planning and General Layout of Heating Plant

11.1.1 Project planning shall be elaborated, including the following:

1 Area of the densified biofuel heating project.

2 Connection of the access roads to the heating plant.

3 Water supply, power supply, heating pipe network routes.

4 The layout of construction site.

5 Design indices for project planning.

11.1.2 Design indices for project planning shall be in accordance with those specified in Table 11.1.2.

Table 11.1.2 Design indices for project planning

S/N	Description	Unit	Value	Remarks
1	Area of heating plant	hm^2		
2	Planned capacity	MW		
3	Project capacity for the current stage	MW		
4	Land occupation per MW	hm^2/MW		
5	Land occupation of main buildings (structures) in heating plant	hm^2		
6	Building factor	%		
7	Utilization area of heating plant	m^2		
8	Utilization factor	%		
9	Excavation/backfill earthwork volume for heating plant	m^3		
10	Area of road and square in heating plant	m^2		
11	Road factor	%		
12	Fencing length for heating plant	m		
13	Green area	m^2		
14	Greening factor	%		

11.1.3 The recommended general layout scheme for the heating plant shall be proposed according to the following requirements:

1 The plan layout and spatial organization of the buildings (structures) in the heating plant shall be compact and reasonable, simple and coordinated, and have clear sectorization, smooth process, safe operation, convenient transport, and easy installation and maintenance.

2 Heating plant planning shall be coordinated with the regional planning; the form and layout of boiler house shall be in harmony with the architectural style of the enterprise and the region.

3 The location of boiler house shall be determined according to the following requirements:

   1) Easy for the centralized conveyance of heating load; the technology for heating pipe laying and outdoor heating pipe network arrangement shall be reliable and cost-effective.

   2) Located in the area with good geological conditions.

   3) Help reduce the impact of dust, harmful gases, noise and ash on the surroundings. Meet the requirements in the environmental impact assessment report.

   4) Easy to store fuel and dispose ash; and the separation of personnel access and fuel, ash transport should be achieved.

   5) In favor of natural ventilation and lighting.

   6) Conducive to the recovery of condensate.

   7) The space for extension should be reserved.

4 The spacing between the boiler house, fuel storage, fuel delivery facilities, ash storage room, offices, living buildings, and their distances from other buildings (structures) shall be specified.

11.1.4 The vertical layout in the plant area shall be put forward.

11.1.5 Construction site and future extension plan shall be briefly described.

11.2 Overall Design of Heating Plant

11.2.1 Fuel handling and storage design shall include:

1 Fuel handling and storage mode and facilities.

2 Type, arrangement and storage capacity of fuel buffer storehouse and long-term storehouse.

3   Damp, rain, and fire prevention measures for fuel.

11.2.2 Fuel delivery form and feeding equipment shall be selected according to fuel characteristics, transport distance, height difference, noise, dust prevention, explosion prevention, fire protection, conveying capacity, investment, operation and maintenance costs, practicality, etc.

11.2.3 Boiler thermal system scheme shall be described, including:

1   Type, model, and number of boilers.

2   Boiler heating medium and parameters.

3   Number of boilers in service in the heating season and backup equipment.

4   Main technological process and equipment parameters of the thermal system.

5   Measures to prevent the boiler from alkali metal corrosion, ash deposit, and clinkering.

6   Measures to control ash and suppress ash corrosion.

11.2.4 Boiler air and flue gas system scheme shall be described, including:

1   Supply fan and induced-draft fan configuration and selection.

2   Flue gas dust removal methods and equipment selection.

3   Flue section, chimney outlet diameter, and height.

11.2.5 Slag removing devices and ash handling shall be explained.

11.2.6 Boiler feed water quality and water treatment scheme shall be described.

11.2.7 The composition and function of the boiler thermotechnical system shall be specified.

11.2.8 The main dimensions of heating boiler house and the layout of main and auxiliary machinery shall be provided.

11.2.9 For retrofitted densified biofuel boiler, the retrofit measures of boiler structure, feeding system, combustion system, air and flue gas system, slag removal equipment and dust removal equipment shall be described.

11.2.10 Bill of quantities (BOQ) for heating plant shall be put forward.

11.3 Overall Design of Heating Pipe Network

11.3.1 The design of heating medium and parameters for a heating pipe network shall include:

1 Requirement and selection of heating medium and its parameters.

2 Heating parameters and mode for the industrial heating load.

3 Heating mode of the space-heating load.

4 Supply mode of heating load for air conditioning.

5 Supply mode of heating load for domestic hot water.

11.3.2 The layout and laying design of heating pipe network shall include:

1 Heating planning on which pipe network route and layout are based.

2 Directions and laying modes of main trunk pipes and branch pipes.

3 Interconnection proposal of a heating system with multi-heat sources.

4 Maximum heating radius and the farthest user distance.

5 Techno-economic comparison of various laying methods.

6 Scheme to allow heating network to span rivers, roads, and railways.

11.3.3 The design of heat supply station shall include:

1 Number, geographical location, heating scale and heating scope of new, renovated or expanded heat supply stations.

2 Connection mode of the heating pipe network, heat supply stations, and heat consuming installations.

3 Selection, quantity, and technical parameters of main equipment in the heat supply station.

11.3.4 The heating regulation and scheduling design for heating pipe network shall include:

1 Heating regulation mode, and operation regulation mode of a heating system with multi-heat sources.

2 Hot water pipe network temperature, water volume regulation data, the temperature curve and water volume curve of heating regulation.

3 Type selection of main equipment and instruments of the system.

11.3.5 Heating metering system scheme shall be specified.

11.3.6 Hydraulic calculation of heating pipe network shall include:

1 Calculation conditions and parameters.

2 Piping diameter.

3   Flow and head of heat source cycle pump and booster pump.

4   Static pressure line and fixed pressure point position.

5   Most unfavorable loop and its water pressure diagram under different operating conditions.

6   Hydraulic calculation table.

11.3.7 Booster pump station process shall be explained, and equipment layout scheme and external supporting facilities shall be put forward, including main equipment model and specification, quantity, technical parameters, booster pump station site selection conditions, electricity consumption, water consumption, power supply, water supply and drainage, etc.

11.3.8 Condensate recovery measures for pipe network shall include:

1   The amount and quality of condensate recovery by each consuming installation, water pressure diagram calculation and condensate pump equipment selection.

2   The reason why the consuming installations cannot retrieve condensate temporarily.

11.3.9 Pipe insulation and anti-corrosion measures shall be put forward according to the different laying methods and their structural requirements of insulation materials.

11.3.10 The scheme to allow the pipeline passing through or spanning roads, railways, rivers, and other obstacles shall be described.

11.3.11 BOQ of heating pipe network shall be put forward.

12 Civil Works, Electrical, Monitoring, Control and Utility System

12.1 Civil Works

12.1.1 The general layout and plan of the heating plant shall be described, and the individual design of main buildings (structures) shall be described.

12.1.2 The main geological conditions of the site area shall be described.

12.1.3 The design of structural type, service life, flood control and waterlogging prevention standards, seismic fortification category, seismic grade of and seismic measures for main buildings (structures) in the heating plant shall be determined. The main structural design for each building shall be described.

12.1.4 The foundation type of main buildings (structures) and ground treatment scheme shall be determined according to the engineering geological conditions in the project area.

12.1.5 Explosion-proof design measures for boiler house shall be defined.

12.1.6 For renovation and extension projects, the reconstruction and utilization of the existing buildings (structures) shall be indicated.

12.1.7 The structural type and the foundation treatment of the structures along the pipeline laying route shall be stated, and the special treatment measures to allow the pipeline passing through or spanning roads, railways, rivers, and other obstacles shall be described.

12.2 Electrical

12.2.1 Main buildings (structures) power load grade, power supply, and voltage shall be determined in accordance with the current national standard GB 50052, *Code for Design Electric Power Supply Systems* according to process requirements, boiler capacity, the importance of heating load, environmental characteristics, building function, etc.

12.2.2 The power supply and distribution system solutions for the heating plant and heating pipe network shall be described.

12.2.3 The illuminance value of artificial lighting for working zones, rooms, and structures shall be determined in accordance with the current national standard, GB 50034, *Standard for Lighting Design of Buildings*.

12.2.4 The technical measures for the lightning protection and earthing of main buildings (structures) shall be put forward.

12.2.5 The BOQ for the electrical system shall be proposed.

12.3 Monitoring and Control

12.3.1 The composition and function of the boiler house's thermal parameter detection and control system, and the main contents of the automatic detection and regulation system shall be described.

12.3.2 The detection and control scheme and functions of the heating system, including heat source, key parts, booster pump station and heat supply station, shall be described.

12.3.3 The structural type and functions of monitoring center and monitoring stations at all levels, main hardware, and software configuration scheme for the overall monitoring and control system of the heating plant and heating pipe network shall be described.

12.3.4 The communication scheme of the heating plant shall be described.

12.3.5 The BOQ for the monitoring and control system shall be proposed.

12.4 Heating, Ventilation, and Air Conditioning

12.4.1 The parameters of indoor air calculation for boiler house, auxiliary equipment room, control room, duty room, office buildings, dormitories, heat supply station, booster pump station, monitoring center and other major production or living rooms shall be determined in accordance with the current national standards GB 50041, *Code for Design of Boiler Plant*, and GB 50019, *Design Code for Heating Ventilation and Air Conditioning of Industrial Buildings*.

12.4.2 The heating design scheme for main production and living rooms shall be briefly described.

12.4.3 The ventilation and air conditioning design scheme for main production and living rooms shall be briefly described.

12.4.4 The BOQ for HVAC system shall be proposed.

12.5 Water Supply and Drainage

12.5.1 The water source solutions for water supply system shall be briefly described.

12.5.2 The water supply system design for main buildings (structures) shall be described.

12.5.3 The drainage system design for sewage, wastewater, and rainwater of main buildings (structures) shall be described, and the necessary measures to be taken shall be explained in accordance with the environmental protection requirements.

12.5.4 The measures to prevent dust blowing in the heating plant, as well as the facilities to prevent ash flushing and stagnant water shall be put forward.

12.5.5 The BOQ for water supply and drainage system shall be proposed.

13 Fire Protection

13.1 Overall Design of Fire Protection

13.1.1 The laws, regulations, and technical standards applicable to the design of fire protection, shall be listed.

13.1.2 Fire protection guidelines and fire protection design principles shall be briefly described.

13.1.3 The fire hazards and their severity in the heating plant shall be analyzed, and the overall fire protection design scheme, involving the functions of firefighting system, public firefighting facilities, fire water sources, fire protection power supply, fire engine lanes, fire escape and firefighting facilities configured to buildings, shall be set forth.

13.2 Fire Protection Engineering Design

13.2.1 The fire separation, fire hazard class, fire-resistance rating and corresponding firefighting measures for main buildings (structures) in the heating plant shall be determined.

13.2.2 The fire protection design of main places and electro-mechanical equipment shall include:

1. The fire protection design and main firefighting facilities for main production sites, offices and living quarters, main electro-mechanical and electrical equipment.

2. The technical measures for critical fire prevention sites and locations such as dumping devices, fuel warehouse, delivery and feeding facilities and boiler slag removal equipment, including firefighting facilities configuration, fire monitoring, fire evacuation routes, emergency lighting, firefighting channels, dust prevention, explosion prevention, etc.

3. The technical measures to prevent stored densified biofuel from spontaneous combustion and being ignited by backfire when feeding.

13.2.3 Firefighting medium design shall be performed as follows:

1. Based on possible fire nature and hazard level, firefighting medium and system firefighting equipment and devices shall be determined.

2. When using water firefighting system, the source of fire water, water supplier, fire water supply capacity and water pressure shall be determined, and the main equipment of fire water pump room and its

layout, water supply pipeline design and distribution of fire hydrants shall be briefly described.

13.2.4 The electrical design for fire protection shall be performed as follows:

    1   The load grade of fire protection power supply, and fire protection power distribution unit.

    2   Fire emergency lighting and fire evacuation indicating signs design.

    3   Automatic fire alarm system, including the composition and configuration scheme and main equipment.

    4   Communication design scheme for fire protection.

13.2.5 The smoke control and exhaust scheme and the fire protection design scheme for heating and ventilation system shall be described.

13.2.6 The fire protection design scheme for architectural decoration and finishing shall be described.

13.2.7 The BOQ for fire protection system shall be proposed.

14 Organization and Staffing

14.1 Project Management Organization

14.1.1 The structure and staffing of the project management organization shall include:

1. Principles for the establishment of the project management organization for heating plant and heating pipe network.

2. Production and operation organization and maintenance organization, as well as the management system and mode for heating plant and heating pipe network.

3. Organizational setup and staffing.

14.1.2 Management scope and regulations shall be determined with consideration of project features.

14.2 Main Management Facilities

14.2.1 Main management facilities shall include:

1. Power supply and its backup required in the project management area.

2. Production and domestic water supply and water supply facilities.

3. Greening plan and area in the project management area.

14.2.2 The transport facilities design for production and living shall be put forward.

15 Project Scheduling

15.1 Project Implementation Conditions

15.1.1 The geographical location and natural conditions shall be outlined.

15.1.2 The accessibility, available land area and conditions of the project site shall be outlined.

15.1.3 The construction conditions, sources and supply conditions of main building materials, construction water and power supply conditions, machining and repair capability, the supply of labor force and supply of living materials in the project area shall be described.

15.1.4 Construction difficulties and characteristics shall be stated.

15.2 Construction Schedule

15.2.1 Construction schedule shall be prepared and shall cover:

1 The principles and basis for determining total construction duration and master construction schedule, the project owner's requirements for the project commissioning date, etc.

2 Construction and installation works, quantities and the key factors affecting the progress of the main works.

15.2.2 The critical path of the master schedule and the factors and conditions controlling the progress of main works shall be stated. The master schedule chart and the intensity indices of main construction items shall be proposed.

16 Environmental Protection, Energy Conservation and Emission Reduction

16.1 Environmental Protection

16.1.1 Environmental protection laws and regulations, policies and technical documents shall be listed, and the environmental protection design standards to follow shall be identified.

16.1.2 The main conclusions of environmental impact statement and the approval opinions of competent authorities shall be stated.

16.1.3 The environmental status shall include:

1. Overview of natural and social environment for the project site, including the project site location, landform, meteorological characteristics, and social environment.

2. Surrounding environment constraints.

3. Local environmental quality and main environmental problems.

16.1.4 The environmental protection objectives of project construction and operation shall be clearly defined.

16.1.5 With respect to the adverse environmental impact of air pollutants, production wastewater, domestic sewage, solid waste, noise and vibration generated by the project, the measures for pollution prevention and control and environmental protection shall be stated for project construction and operation periods, including flue gas purification, wastewater discharge control, solid waste management, noise and vibration control, and greening in plant area.

16.1.6 The cost estimate for environmental protection measures shall be proposed.

16.1.7 The assessment and conclusion on environmental benefits and environmental protection design of the project shall be put forward.

16.2 Energy Conservation and Emission Reduction

16.2.1 The laws and regulations and technical standards to follow in the design of energy conservation and emission reduction shall be listed.

16.2.2 The type of energy consumption, energy consumption indicators and energy consumption during operation period shall be analyzed as follows:

1. The main energy-consuming equipment of heating system shall be analyzed according to general project layout, and the annual energy

consumption and the corresponding energy efficiency indicators of the heating system shall be proposed.

2　According to the type, size and functional requirements of main buildings (structures) in heating plant, and the design schemes of HVAC, electrical, lighting, and water supply and drainage, the energy consumption of buildings (structures) shall be explained, and annual energy consumption and the corresponding energy efficiency indicators shall be proposed.

3　Energy consumption and main energy-consuming equipment and facilities during operation period shall be comprehensively analyzed, the control indicators of energy consumption during operation period shall be proposed, including service power consumption indicator and total power consumption indicator for heating plant and heating pipe network.

16.2.3 The main energy-saving measures shall be described, including:

1　The energy-saving factors in the selection and layout of the main equipment for heating projects, as well as appropriate measures; the energy-saving factors in the design of the main buildings (structures), as well as appropriate measures.

2　The implementation of mandatory energy-saving standards in the design of boiler house auxiliary machine system, HVAC, electrical, lighting, and water supply and drainage, as well as appropriate measures and their effectiveness.

3　The suggestion of energy-saving measures during operation period.

16.2.4 Energy conservation and emission reduction benefits shall be analyzed as follows:

1　Calculation of the project heat output, and the equivalent heating capacity of a fossil-fueled project.

2　The energy-saving benefits such as the calculation result indicating how much fossil energy can be saved shall be described according to the heating system structure and its utilization efficiency in the heating service area of the project.

3　The total emission reduction amount of greenhouse gas and other pollutants shall be described, and the benefits of greenhouse gas and other pollutant emission reduction shall be analyzed according to fossil energy savings.

17 Occupational Health and Safety

17.1 Identification and Analysis of Main Hazard Factors

17.1.1 The possible unsafe factors and hazards in project site selection and its general layout that arise from geological and natural conditions and the social environmental conditions in surrounding areas shall be analyzed.

17.1.2 The possible hazards in the production process, such as fire, explosion, mechanical injury, object strike injury, fall injury, burning injury, lifting injury, vehicle injury, natural disasters, as well as possible consequences of serious personal injury or death and property damage, shall be analyzed.

17.1.3 The harmful factors such as noise, vibration, high temperature, dust, poisonous and harmful substances that may exist in the production sites of plant area and the serious consequences that may endanger the physical and mental health of staff shall be described.

17.1.4 Identification and analysis of major hazard sources of the project shall be conducted.

17.2 Engineering Design of Occupational Health and Safety

17.2.1 With respect to the possible unsafe factors in project site selection and its general layout that arise from geological and natural conditions and the social environmental conditions in surrounding areas, safety measures shall be proposed.

17.2.2 The safety measures in terms of fire separation, fire engine lane, evacuation routes, and safety clearances for buildings (structures) in the heating plant shall be proposed.

17.2.3 Specific safety precautions shall be put forward based on the analysis of the main hazards in the production and project features.

17.2.4 Specific occupational health precautions shall be put forward based on the analysis of the harmful factors in main production workplaces and project features.

17.2.5 The locations where safety signs need to be provided shall be listed, and requirements for the safety sign type, graphics, text, and color shall be described.

17.2.6 The setting and technical requirements of auxiliary room for safety and health shall be put forward; the plant safety management organization and full-time and part-time management staffing shall be specified; the configuration of safety and health instruments and equipment for education and training shall be determined.

17.3 Safety Management During Operation

17.3.1 The safety management scope during operation period shall be determined according to the actual situation.

17.3.2 The requirements of safety management system for heating plant shall be proposed.

17.3.3 The purpose, requirements and main contents of emergency response plan shall be stated, and the main items of the plan shall be put forward.

17.4 Special Investment for Occupational Health and Safety

17.4.1 The principles, basis and price level year for special investment preparation shall be specified.

17.4.2 The cost estimate for occupational health and safety measures shall be proposed.

18 Cost Estimate

18.1 Preparation Instructions

18.1.1 The main materials consumption, construction schedule, funding sources, and financing structure of the project shall be outlined; the total project investment, static investment, and investment per MW heat output shall be described.

18.1.2 The preparation basis and principles shall be proposed according to the following requirements:

1 The provisions on estimate preparation, quota and cost standards, design documents, the price level year for static investment preparation, etc. shall be specified.

2 Cost estimate shall be prepared in accordance with GB 50500, *Code of Valuation with Bill Quantity of Construction Works* and the regulations and cost standards for the consumption quota for local construction and installation works.

18.1.3 The principles and basis shall be determined according to the principles and basis for calculating the unit price of labor budget, the price of main materials, and other basic prices, the original prices and the transportation mode of equipment, and the basic price shall be specified.

18.1.4 The rate standards shall be specified according to the rate indices adopted in calculating the unit price for equipment installation and construction works, basic contingency rate, loan interest rate, exchange rate, and other cost calculation standards.

18.2 Cost Estimate Table

18.2.1 Cost estimate tables shall be prepared, including summary cost estimate, equipment and installation work cost estimate, construction works cost estimate, other cost estimates, and annual cost allocation.

18.2.2 The appendix for budget estimate shall include the budgetary price calculation for main materials, the machine-hour calculation for main construction machinery, the unit price summary for installation works, the unit price summary for construction works, the unit price calculation for the project, and the documents and price information related to the budgetary price and cost calculation for labor, materials and equipment.

19 Financial Evaluation and Social Benefit Analysis

19.1 Overview

The project scale, annual average sales of heat, construction period and financial evaluation period shall be briefly described, and the financial evaluation basis shall be provided.

19.2 Financial Evaluation

19.2.1 The construction fund composition of the project shall be briefly described, including investment in fixed assets, interest incurred during construction, and liquidity. Funding options and loan repayment conditions shall be explained.

19.2.2 Financial analysis and evaluation shall include:

1. Total cost calculation, including fixed asset value and liquidity calculation; total project cost calculation, including operating costs, depreciation charges, amortization charges, and interest expenses, of which operating costs include maintenance fees, employee wages and welfare, insurance premiums, material costs, and other expenses.

2. Heat supply benefits calculation, involving the calculation methods and parameters of heat supply benefits, including year-by-year heat supply revenue, input tax credit, profit and distribution, tax calculation.

3. Liquidity analysis, involving the calculation of loan repayment with interest and assets and liabilities, the analysis of solvency, and the proposal of interest reserve ratio, debt service ratio and asset-liability ratio.

4. Profitability analysis, involving project financial cash flow calculation, equity financial cash flow calculation; the analysis of financial evaluation indicators such as FIRR before and after income tax, return on investment, investment profit, and tax rate, and return on equity based on financial profitability calculation results.

5. Financial viability analysis, mainly involving financial plan cash flow calculation.

6. Sensitivity analysis. The uncertainties of densified biofuel heating projects mainly include fuel cost, sales of heat, investment in fixed assets, sales price, etc. The changes in FIRR arising from the change in the above factors shall be calculated; the impact of the changes in

fuel costs, sales of heat, and investment in fixed assets on the sale price of heat shall be calculated on a benchmark yield basis, to analyze the project's ability to resist risks.

7   Uncertainty analysis.

19.2.3 A summary table of financial evaluation indicators shall be prepared and the financial feasibility evaluation conclusions for the project shall be presented.

19.3   Social Benefit Evaluation

19.3.1 Analysis and evaluation shall focus on the practical and long-term benefits of the project for urbanization, atmospheric environment, people's livelihood, income of local farmers, renewable energy development, etc.

19.3.2 The social benefits in energy conservation and emission reduction of the project shall be described.

19.4   Financial Evaluation Table

19.4.1 Financial evaluation tables shall be provided, including the investment plan and fundraising table, total cost table, profit, and profit distribution table, project investment cash flow statement, equity cash flow statement, financial plan cash flow statement, repayment expense statement, a summary table of financial indicators, etc.

20 Project Bidding

20.1 Bidding Scope

The bidding scope and main lots for the project shall be specified.

20.2 Bidding Method and Organization

The bidding method and organization shall be determined, and a basic bidding information table shall be prepared. Basic information for bidding shall be in accordance with Table 20.2.1.

Table 20.2.1 Basic information for bidding

Project:

	Bidding scope		Organization		Bidding method		Not for bidding	Estimated bid amount (CNY)	Remarks
	All works	Partial works	By own	By agency	Competitive	Selective			
Investigation									
Design									
Construction									
Installation									
Supervision									
Equipment									
Important material									
Others									
Further information: Seal of the owner Date:									

NOTE Further information may be continued on the next page.

21 Conclusions and Suggestions

21.1 Conclusions

21.1.1 The main content of the recommended project plan shall be described.

21.1.2 The main conclusions of economic analysis and social benefits shall be provided.

21.2 Suggestions

21.2.1 The main issues to be solved and suggestions shall be put forward.

22 Attachments and Drawings

22.1 Attachments

22.1.1 The feasibility study report on densified biofuel heating projects should include the following attachments:

1. Project approval documents of relevant competent authorities.
2. Approval documents of local planning management departments.
3. Resource survey and evaluation report of agriculture and forest biomass heating projects.
4. Agreement of intent for supplying densified biofuel.
5. Agreement of intent for heat supply with users.
6. Approval documents on the project environmental impact statement.
7. Other supporting documents.

22.2 Drawings

22.2.1 The feasibility study report on densified biofuel heating project may include the following drawings:

1. Local heating plan chart.
2. Local densified biofuel resources and supply sources distribution map.
3. Geographical location map of densified biofuel heating project.
4. General plan of heating plant area.
5. Vertical layout of heating plant area.
6. Boiler house layout plan.
7. General piping plan in the heating plant area.
8. Thermodynamic system drawing, water treatment system drawing, flue gas purification system drawing of boiler house.
9. Fuel delivery and feeding system layout for the boiler house.
10. Single line diagram, power supply and distribution system diagram for the boiler house.
11. Transformer and distribution equipment layout for the boiler house.
12. Plan, elevation and profile of the boiler house building.
13. Thermotechnical test and control system schematic for the boiler house.

14 Heating load diagram.

15 General layout for heating pipe network.

16 Process flow chart of heat supply station and booster pump station.

17 Layout plan of booster pump station process equipment.

18 Monitoring and control system diagram for heating pipe network.

19 Project construction land boundary map.

20 Other necessary schematic layouts.

Appendix A Basic Data for Preparation of Feasibility Study Report on Densified Biofuel Heating Projects

A.0.1 Basic data on heating plant shall meet the following requirements:

1 Heating load data, which shall include:

1) Heating medium and parameter requirements.

2) Hourly maximum and hourly average heat demands required by production, heating, ventilation, and living.

3) Maximum and average hourly heat production for residual heat utilization, and steam or hot water parameters.

4) Collaborative heating data of neighboring firms, including heat source delivery distance, heating load, medium parameters, price, condensate return water requirements, etc.

5) Heat utilization trend of the user, including whether there is a plan for extension in phases, increase in heating load, or whether there are heating plants nearby, whether there is a possibility to use other heat supply.

6) Condensate recovery amount.

7) Heating load curve for the whole plant or district.

2 Densified biofuel analysis data, which shall include:

1) Fuel type: densified biofuel may be classified according to shape (pellets, briquettes, and logs) and raw materials (herbs, woods, and others).

2) Total moisture.

3) Proximate analysis (generally including moisture, volatile matter, ash content, fixed carbon).

4) Ultimate analysis (carbon, hydrogen, oxygen, nitrogen, sulfur).

5) Calorific value.

6) Specification.

7) Bulk density.

8) Density.

9) Mechanical durability.

3 Water source and water quality data. Water source data shall include the water source condition for water supply at the plant site. Water quality data shall include suspended solids, dissolved solids, total hardness, carbonate hardness, non-carbonate hardness, calcium hardness, magnesium hardness, total alkalinity, oxygen consumption, oil content, free carbon dioxide, dissolved oxygen, pH, etc.

4 Meteorological data items shall be in accordance with Table A.0.1.

Table A.0.1   Meteorological data items

S/N	Item		Unit	Value
1	Altitude		m	
2	Outdoor design temperature	Winter heating	°C	
		Winter ventilation	°C	
		Summer ventilation	°C	
3	Outdoor mean temperature during heating period		°C	
	Number of design days used in heating period		d	
4	Dominant wind direction and its frequency	In winter		
		In summer		
5	Atmospheric pressure	In winter	kPa	
		In summer	kPa	
6	Maximum depth of frozen ground		cm	

5 Geological data shall include:

  1) Collapsible loess, groundwater table, ground bearing capacity, etc.

  2) Basic seismic intensity and peak ground acceleration.

6 The topographical map of heating plant shall be collected.

7 Equipment and material data shall include:

  1) Boiler unit data: the main technical parameters, models, specifications, outline drawings and prices of the boiler and auxiliary machinery.

2) Auxiliary machine equipment data: fans, pumps, various standard and non-standard equipment drawings, technical parameters, and prices.

3) Materials: local insulation materials, pipes, steel, etc.

A.0.2 In addition to basic data, the following data shall be collected for renovation or extension heating plant.

1 The models, specification, quantity, manufacturers, service life, main dimensions, operating conditions and problems for existing boiler house and equipment.

2 The as-built drawing for existing boiler house.

3 Differences between the actual situation on the site and as-built drawing.

4 The operation record, problems and accident analysis of existing boiler house.

5 The staffing of the existing boiler house.

A.0.3 The basic data of district heating pipe networks shall include:

1 1 : 1 000 topographic map for the heating area.

2 Local meteorological data:

1) Outdoor design temperature for winter heating (°C).

2) Outdoor design temperature for winter ventilation (°C).

3) Annual mean monthly temperature (°C).

4) Number of days with daily mean temperature not exceeding 5 °C (d).

5) Mean temperature in the period with daily mean temperature not exceeding 5 °C (°C).

6) Extreme low temperature (°C).

7) Outdoor mean wind speed in winter (m/s).

8) Outdoor mean dominant wind direction and frequency in winter.

9) Wind rose diagram.

10) Maximum depth of frozen ground (cm).

11) Atmospheric pressure in winter (kPa).

3 Hydrogeological investigation report: soil properties, soil pressure

resistance, groundwater table.

4 Heating load for production, HVAC, refrigeration, hot water supply.

5 Short-term and long-term development planning and status of development year by year for the heating area.

Appendix B Contents of Preparation of Feasibility Study Report on Densified Biofuel Heating Projects

1　Executive Summary

2　Necessity and Feasibility of Construction

2.1　Project background

2.2　Necessity of construction

2.3　Feasibility of construction

3　Heating Load

3.1　Present heating status

3.2　Heating load

3.3　Design heating load

3.4　Heating load diagram

4　Densified Biofuel Supplies

4.1　Market demand analysis

4.2　Supply and security

5　Project Scale

6　Properties of Densified Biofuel and Boiler Selection

6.1　Properties of densified biofuel

6.2　Boiler selection

7　Project Site Selection

7.1　Site selection

7.2　Site overview

7.3　Engineering geology

7.4　Traffic conditions

7.5　Water and power supply

8　General Scheme

8.1　Project planning and general layout of heating plant

8.2　Overall design of heating plant

8.3 Overall design of heating pipe network

9 Civil Works, Electrical, Monitoring, Control and Utility System

9.1 Civil works

9.2 Electrical

9.3 Monitoring and control

9.4 Heating, ventilation, and air conditioning

9.5 Water supply and drainage

10 Fire Protection

10.1 Overall design of fire protection

10.2 Fire protection engineering design

11 Organization and Staffing

11.1 Project management organization

11.2 Main processing facilities

12 Project Scheduling

12.1 Project implementation conditions

12.2 Construction schedule

13 Environmental Protection, Energy Conservation and Emission Reduction

13.1 Environmental protection

13.2 Energy conservation and emission reduction

14 Occupational Health and Safety

14.1 Identification and analysis of main hazard factors

14.2 Engineering design of occupational health and safety

14.3 Safety management during operation

14.4 Special investment for occupational health and safety

15 Cost Estimate

15.1 Preparation instructions

15.2 Cost estimate table

16 Financial Evaluation and Social Benefit Analysis
16.1 Overview
16.2 Financial evaluation
16.3 Social benefit evaluation
16.4 Financial evaluation table

17 Project Bidding
17.1 Bidding scope
17.2 Bidding method and organization

18 Conclusions and Suggestions
18.1 Conclusions
18.2 Suggestions

19 Attachments and Drawings
19.1 Attachments
19.2 Drawings

Appendix C Main Techno-Economic Index of Densified Biofuel Heating Projects

C.0.1 Main techno-economic indices of boiler house in heating plant shall be in accordance with Table C.0.1.

Table C.0.1 Main techno-economic indices of boiler house in heating plant

S/N	Description			Unit	Value	Remarks
1	Heat output of boiler house	Max. heating load	Heating period	GJ/h (t/h)		
			Non-heating period	GJ/h (t/h)		
		Mean heating load	Heating period	GJ/h (t/h)		
			Non-heating period	GJ/h (t/h)		
		Min. heating load	Heating period	GJ/h (t/h)		
			Non-heating period	GJ/h (t/h)		
2	Boiler thermal efficiency			%		
3	Area	Floor area of boiler house		m^2		
		Land use area for boiler house		m^2		
4	Annual heat output	Steam		t/a		
		Hot water		GJ/a		
5	Heating radius	Steam		km		
		Hot water		km		
6	Heating medium parameters	Supply water temperature		°C		
		Return water temperature		°C		
		Steam pressure		MPa		
7	Total investment			CNY		
8	Unit investment	By heating load		CNY/GJ		
		By steam quantity		CNY/t		

Table C.0.1 *(continued)*

S/N	Description		Unit	Value	Remarks
9	Fuel consumption rate	Annual average heating fuel consumption rate for steam	kg/t		
		Annual average heating fuel consumption rate for hot water	kg/GJ		
10	Staffing		person		
11	Unit cost	Steam	CNY/t		
		Hot water	CNY/GJ		
12	Sales price	Steam	CNY/t		
		Hot water	CNY/GJ		
13	Energy consumption index for boiler house	Water Consumption	kg/t, kg/MW		
		Power consumption	kWh/t, kWh/MW		
		Fuel consumption	kg/t, kg/MW		
14	Annual furnace ash		t/a		
15	Annual water consumption		t/a		
16	Annual power consumption		kWh/a		
17	Annual fuel consumption		t/a		

C.0.2 Main techno-economic indices of heating pipe network shall be in accordance with Table C.0.2.

Table C.0.2 Main techno-economic indices of heating pipe network

S/N	Description			Unit	Value	Remarks
1	Heat output	Max. heat output	Heating period	GJ/h		
			Non-heating period	GJ/h		
		Mean heat output	Heating period	GJ/h		
			Non-heating period	GJ/h		
		Min. heat output	Heating period	GJ/h		
			Non-heating period	GJ/h		

Table C.0.2 *(continued)*

S/N	Description		Unit	Value	Remarks
2	Heating area	Industrial building	m²		
		Public building	m²		
		Residential building	m²		
3	Annual heat output	Steam	t/a		
		Hot water	GJ/a		
4	Heating radius	Steam	km		
		Hot water	km		
5	Heating medium parameters	Supply water temperature	°C		
		Return water temperature	°C		
		Steam pressure	MPa		
6	Average pressure drop of pipeline	Steam	MPa/km		
		Hot water	MPa/km		
7	Heating load density	Heating load for industry	GJ/km²		
		Space-heating load	GJ/km²		
		Heating load for cooling by air conditioner	GJ/km²		
		Heating load for hot water supply	GJ/km²		
8	Total investment		CNY		
9	Unit investment	By heating load	CNY/GJ		
		By heating distance	CNY/km		
10	Equivalent to standard coal consumption	Annual average heating for steam	kg/t		
		Annual average heating for hot water	kg/GJ		
11	Staffing		Person		
12	Unit cost	Steam	CNY/t		
		Hot water	CNY/GJ		
13	Sales price	Steam	CNY/t		
		Hot water	CNY/GJ		

Table C.0.2 *(continued)*

S/N	Description		Unit	Value	Remarks
14	Internal rate of return		%		
15	Return on investment		%		
16	Return on equity		%		
17	Annual saving of standard coal		t		
18	Environmental protection benefits	Dust emission reduction	t/a		
		SO_2 emission reduction	t/a		
		NO_x emission reduction	t/a		
		CO_2 emission reduction	t/a		

Explanation of Wording in This Specification

1. Words used for different degrees of strictness are explained as follows in order to mark the differences in executing the requirements in this specification.

    1) Words denoting a very strict or mandatory requirement:

        "Must" is used for affirmation; "must not" for negation.

    2) Words denoting a strict requirement under normal conditions:

        "Shall" is used for affirmation; "shall not" for negation.

    3) Words denoting a permission of a slight choice or an indication of the most suitable choice when conditions permit:

        "Should" is used for affirmation; "should not" for negation.

    4) "May" is used to express the option available, sometimes with the conditional permit.

2. "Shall meet the requirements of…" or "shall comply with…" is used in this specification to indicate that it is necessary to comply with the requirements stipulated in other relative standards and codes.

List of Quoted Standards

GB 50019,	*Design Code for Heating Ventilation and Air Conditioning of Industrial Buildings*
GB 50034,	*Standard for Lighting Design of Buildings*
GB 50041,	*Code for Design of Boiler Plant*
GB 50052,	*Code for Design Electric Power Supply Systems*
GB 50500,	*Code of Valuation with Bill Quantity of Construction Works*
GB/T 15317,	*Monitoring and Testing for Energy Saving of Coal Fired Industrial Boilers*
GB 18306,	*Seismic Ground Motion Parameters Zonation Map of China*
GB 24500,	*The Minimum Allowable Values of Energy Efficiency and Energy Efficiency Grades of Industrial Boilers*